EXPLORING CAREERS WITHOUT COLLEGE

VOCATIONAL CAREERS IN ENGINEERING

by Cynthia Kennedy Henzel

BrightPoint Press
San Diego, CA

BrightPoint Press

an imprint of ReferencePoint Press, Inc.
Printed in the United States

For more information, contact:
BrightPoint Press
PO Box 27779
San Diego, CA 92198
www.BrightPointPress.com

LIBRARY OF CONGRESS CATALOGING-IN-PUBLICATION DATA

Name: Henzel, Cynthia Kennedy, author.
Title: Vocational careers in engineering / by Cynthia Kennedy Henzel.
Description: San Diego, CA: ReferencePoint Press, 2026 | Series: Exploring careers without college | Includes bibliographical references and index. | Audience: Grades 7–9
Identifiers: ISBN 9781678212728 (hardcover) | ISBN 9781678212735 (eBook)
The complete Library of Congress record is available at www.loc.gov.

CONTENTS

THE ENGINEERING INDUSTRY AT A GLANCE

ANNUAL MEAN WAGE OF ENGINEERING TECHNOLOGISTS AND TECHNICIANS, 2023

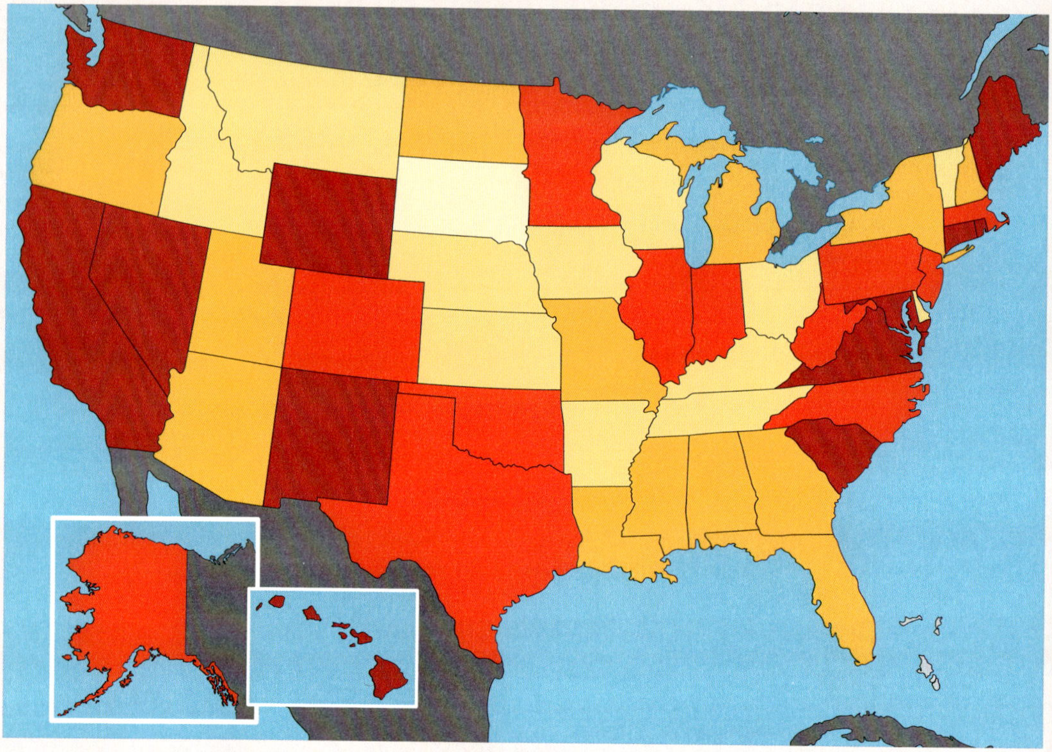

No data

$42,830 to $63,960

$64,150 to $69,800

$70,360 to $79,620

$79,780 to $105,390

Source: "Engineering Technologists and Technicians, Except Drafters, All Other," US Bureau of Labor Statistics, April 3, 2024, *www.bls.gov.*

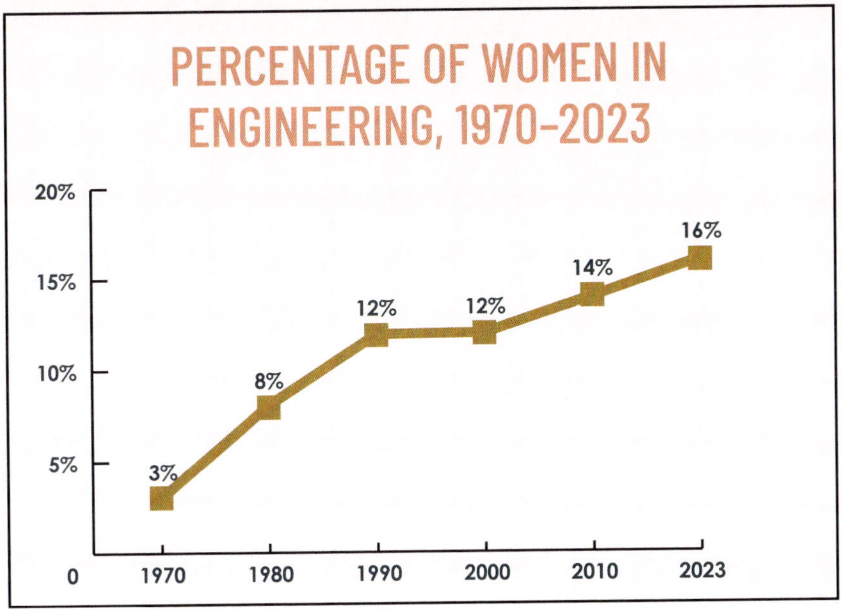

Source: "Percentage of Women Workers in Science, Technology, Engineering, and Math (STEM)," US Department of Labor, April 2025, www.dol.gov.

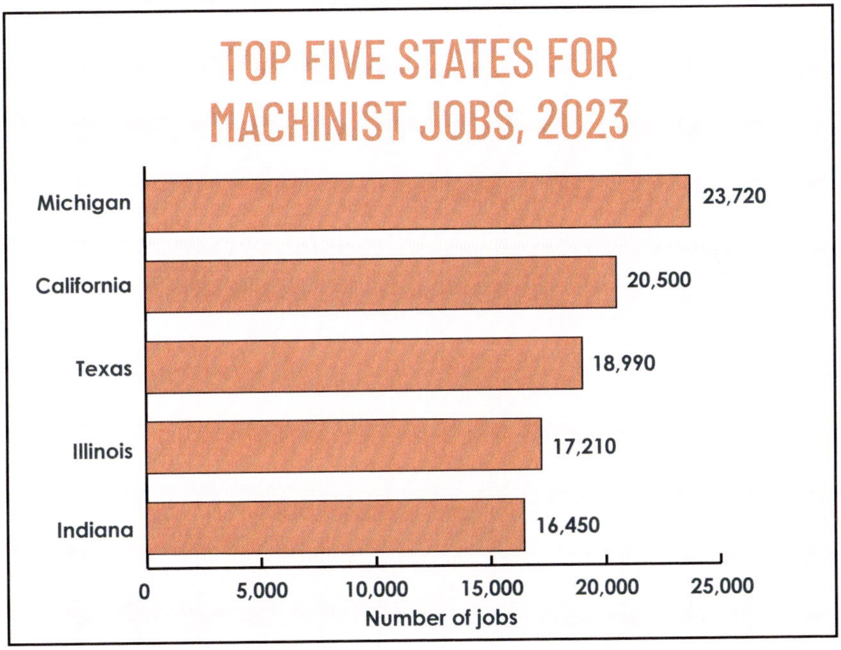

Source: "Occupational Employment and Wage Statistics: Machinists," US Bureau of Labor Statistics, April 3, 2024, www.bls.gov.

WHAT IS THE ENGINEERING INDUSTRY?

Steven Utenhagen was an environmental engineering technician. One day, Utenhagen was at a gold mine. He hiked up to a pipe. Water from the mine flowed from the pipe to a treatment wetland. Treatment wetlands help take out **contaminants** from water. They ensure water is safe before it goes back into the environment.

Utenhagen filled two bottles with water from the pipe. He sealed them. They were

Taking samples is an important part of an environmental engineering technician's job.

Industrial waste, oil, and plastic are some of the many pollutants that can harm the environment.

labeled with the date and place. Utenhagen then went back to the mine's lab. He tested the samples for chemicals. The water would be turned off if he found anything. Then he would help investigate where the pollution came from.

That day, the samples were normal. But his tests were just the first step. He packaged them to send to a larger lab.

That lab had more equipment. It would do more tests. Utenhagen enjoyed his job. He played an important role in preventing contamination in the environment.

WORKING IN THE ENGINEERING INDUSTRY

Engineers find solutions to everyday problems. They use math and science to create and design products. They work in a variety of fields. These could be related to space, technology, or the environment.

Engineers make sure that different structures are safe. They help create more effective medicines. And they develop better machines for manufacturing. Whenever there is a problem to solve, engineers can help solve it.

CHAPTER ONE

ENVIRONMENTAL ENGINEERING TECHNICIAN

Environmental engineering technicians work with environmental engineers. They carry out the plans made by engineers. Their job is to control damage caused by environmental problems. These include pollution. Techs also work to prevent pollution. In the field, they set up equipment that monitors the environment. They collect samples of water, soil, and air.

Environmental techs often check the air quality inside or outside places such as factories, industrial plants, and mines to ensure these sites meet government and safety laws.

Environmental techs run tests for **hazardous** chemicals. This is done in a lab. Techs must accurately record and report their findings. The reports are used

Environmental Engineering Technician

Minimum Education: High school diploma, trade school
Personal Qualities: Enjoys working outdoors, good communication and critical-thinking skills
Certification and Licensing: Certifications and licenses sometimes required
Working Conditions: Environmental engineering technicians work in the field or in labs. They may have to deal with bad weather or different land features.
Average Salary: $58,890 (2024)
Number of Jobs: 14,200 (2023)
Future Job Outlook: 2 percent growth expected between 2023 and 2033

by the US government to create laws that help protect the environment. Different industries must follow these laws.

Techs also keep track of equipment and materials. They must have good communication skills. Techs often work on a team. They must report observations and problems to an environmental engineer. Problem-solving skills are also important. Techs use their skills to look for answers to unexpected findings. They need to understand legal and technical documents, too.

GETTING STARTED

Environmental techs need a high school diploma. Students should take science and math courses. Some students might work in the field to gain experience. For example, they might get a job at a wastewater treatment plant. Joe Proulx is an engineering technology instructor. He works at a college. Proulx says, "Many students at [my] school are already working in the field when they enroll."[1] They go to school to advance in their careers.

Some environmental tech students might do an internship. Internships help students gain experience and learn skills from professionals in the field.

Other students enroll in environmental tech programs right after high school. These 2-year programs are taught at junior colleges or trade schools. Students learn chemistry and math. Some programs have courses in environmental assessment. This means students learn how to collect information about an area. Engineers use the information to decide how a new project might affect a location. Students also learn to deal with hazardous waste.

Some states require techs to have a license to remove hazardous waste. Techs may also need certification. This is a document proving a person has certain knowledge or skills. One certification is the Hazardous Waste Operations and Emergency Response certificate. It requires training in identifying hazards. More training programs are found on the Occupational Safety and Health Administration website.

ON THE JOB

Environmental techs do many tasks. They help engineers check and maintain equipment. This equipment is used to prevent pollution. It also helps clean up pollution. Techs set up and operate equipment in the field. They adjust it for a particular site. They may have to work around weather or different land features.

Other techs monitor industries. They collect air and water samples near factory sites. The techs maintain records. They make sure reporting requirements are met. Some sites are permanently closed.

Portable water quality meters allow environmental engineering techs to test water in the field. This helps techs collect data and analyze water-sample results quickly.

Techs make sure contaminants left behind do not affect people in the area.

Some techs inspect old buildings. They test for **toxic** chemicals. One of these is lead paint. Lead causes serious health problems. Today, it is illegal to use lead in paint. But some older homes may still have paint containing lead.

Another toxic material found in homes is asbestos. Asbestos is a mineral fiber. It was commonly used in construction. This is because it is fireproof. Asbestos is found in floor tile and siding. It was used to protect walls and pipes from heat or cold. However, breathing it in can cause lung diseases. The material is now banned. But it remains

Environmental Services Technicians

Environmental services technicians (ESTs) often work in medical buildings. ESTs help prevent diseases and infections from spreading. They receive specialized training in cleaning rooms and equipment. They have to meet safety and government regulations. ESTs also learn about new technology. This includes systems that help destroy germs.

in some older homes. Techs must know how to properly throw away these materials.

Sometimes, techs respond to emergencies. These might be oil spills. They might involve hazardous chemicals dumped into a waterway. Or they could be radioactive waste dumped during a train crash. Techs work to keep hazardous materials from spreading. They make sure the spill is contained. Then they continue observing the site. Cleaning up these accidents often takes years.

Techs may be exposed to hazardous materials. They wear special suits to protect themselves. These suits are made of rubberlike materials. They cover the body from head to foot. Sometimes, techs wear respirators. Respirators are masks that filter

Environmental techs must know how to properly clean up and get rid of hazardous materials.

the air. They remove toxic chemicals. When working in water, techs wear heavy rubber boots. This helps keep their feet dry.

Proulx says environmental engineering techs must make sure people follow safety rules. He notes that can come with challenges. But the job is still rewarding. "You're doing the right thing to protect the environment," Proulx explains.[2]

FIND OUT MORE

Environmental Sampling and Analytical Methods Program

www.epa.gov/esam

The Environmental Protection Agency's Environmental Sampling and Analytical Methods (ESAM) program offers information and videos on the proper ways to deal with contamination emergencies.

National Institute for Certification in Engineering Technologies

www.nicet.org

The National Institute for Certification in Engineering Technologies (NICET) is an organization that provides training and exams in different types of engineering technology.

Occupational Safety and Health Administration

www.osha.gov

The Occupational Safety and Health Administration (OSHA) site provides information on OSHA certification requirements and training. It also includes information about safety regulations.

CHAPTER TWO

CAD TECHNICIAN

Computer-aided design (CAD) technicians use CAD software. This software helps people make technical drawings. CAD techs work with engineers to design parts used in manufacturing. They work with architects to design buildings. CAD designs are used to estimate the cost for new projects. Workers then use the designs as plans to create the products.

CAD technicians are sometimes called CAD drafters or operators.

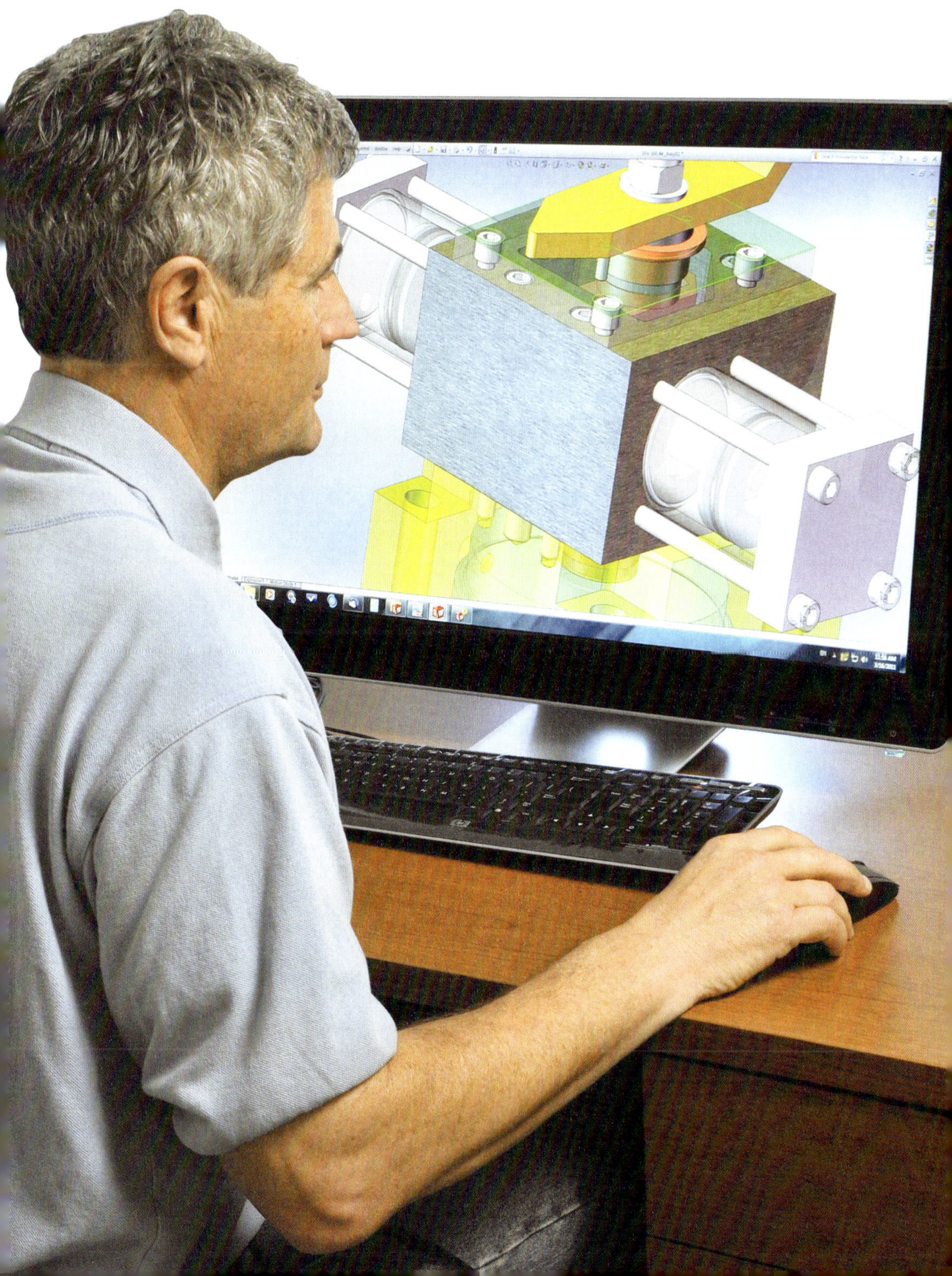

CAD technicians create two-dimensional (2D) designs. Designs can also be three-dimensional (3D). This allows manufacturers to see an object from different directions. Using 3D models lets

CAD Technician

Minimum Education: High school diploma, trade school

Personal Qualities: Strong math and computer skills, ability to work on a team, detail-oriented

Certification and Licensing: Certifications sometimes required, such as the CAD Technician Certificate

Working Conditions: CAD technicians are usually in an office working on computers. Techs might meet with others to discuss projects or go visit sites in person.

Average Salary: $65,380 (2024)

Number of Jobs: 204,500 (2023)

Future Job Outlook: 1 percent decline expected between 2023 and 2033

CAD techs work in many industries, including manufacturing, construction, and transportation.

manufacturers make design changes before creating a **prototype**. Many industries use CAD. For example, architects use models to view structures. Or they can view entire neighborhoods.

Technicians create scale models. These allow architects to make sure buildings will properly function in the areas they are built. CAD techs might also work with

AutoCAD, SolidWorks, and Fusion 360 are just a few of the many programs that CAD techs might use.

product designers. They can make design changes much faster than with traditional models. These include models for new car designs.

CAD techs help engineers see how mechanical parts work. They ensure different parts and gears work together. CAD is even used in filming movies.

Technicians use CAD software to create animation and special effects. They might also use it for virtual reality computer games.

GETTING STARTED

CAD technicians need a high school diploma. Math skills are important, especially geometry. Industries that use CAD require different skills. For example, art classes are helpful for CAD techs working in interior design. Physics classes might be useful for those working in engineering design.

Felicia Bennett is a CAD technician. She explains that being a CAD tech never has a dull moment. "There is always something different that gives you challenges, and it's very interesting to try and decide how to overcome this obstacle," she says.[3]

Many companies offer on-the-job training for CAD techs. Trade schools offer certifications in CAD technology. Certifications are usually earned for knowing how to use a particular software program. Some companies require a CAD Technician Certificate.

AutoCAD is a basic program used in many industries. But there are also special

CAD in Medicine

CAD technicians play an important role in medicine. They create tools for surgery. They also design implants and artificial limbs. CAD techs might design artificial parts for hearts. These can be custom made for a patient. Some CAD programs even design dental products.

CAD software programs. For example, Altium is used for designing electrical circuits. Students can learn to use CAD software online. FreeCAD is available online at no cost. Many CAD programs are similar. If someone knows one program, it is easier to learn another.

ON THE JOB

CAD techs often work with a team. Team members may include engineers or surveyors. They might also include designers. The team works toward a common goal. Good communication skills are important. Techs must understand what is needed by different team members. They must also communicate their own ideas and concerns.

Most CAD technicians work 40-hour weeks. However, they might work more hours to meet a deadline. Techs might be working on several projects at once. They have to decide what needs their attention first. Some CAD techs work from home. Others work in an office.

Techs spend most of their days creating and looking over designs. Designs may need to be updated with new engineering ideas. The CAD tech checks that the measurements are accurate. They might meet with their team. The team members share what they are doing. This helps the tech know whether they need to adjust the design or schedule.

CAD techs might also visit sites in person. They may take measurements of

Working on a team gives CAD techs the opportunity to receive feedback or hear new ideas that can help improve designs.

products in a factory. Doing this ensures the end product meets the **specifications** of the design. CAD techs might also visit building sites. Techs check that their design matches a structure or product in person. They report on any issues they find in the designs. Or they make notes on any changes that need to be made.

Heather Walk makes building designs with CAD. She says, "I like watching what's put on the computer come to **fruition**. Being able to see it physically and know I was a part of creating it—that, for me, is the best."[4]

FIND OUT MORE

American Design Drafting Association
www.adda.org
The American Design Drafting Association (ADDA) is a professional organization that provides information about educational and employment opportunities for CAD techs. ADDA also sponsors poster and design drafting competitions.

Association for Computer Aided Design in Architecture
www.acadia.org
The Association for Computer Aided Design in Architecture (ACADIA) site includes examples of computer-aided designs. ACADIA also features important news in the field and has information about events and workshops people can participate in.

CHAPTER THREE

SURVEYING AND MAPPING TECHNICIAN

Surveying and mapping technicians support surveyors. Survey techs help collect data about an area of land. They measure exactly where the land is located on Earth. Techs find the boundaries of a property. They measure **elevation**. They also record special features. These include streams and roads.

Survey techs then use the information to create maps. Maps are used by

Surveying and mapping technicians work outside year-round. Some tasks may be easier or harder to do depending on the weather.

city planners. They need information about property lines and **zoning**. Utility companies use maps to find existing pipes. Builders use maps to create new communities. Survey techs place markers. These show where a large property will

Surveying and Mapping Technician

Minimum Education: High school diploma
Personal Qualities: Detail-oriented, physically fit
Certification and Licensing: Not required, but certifications can help advancement
Working Conditions: Surveying and mapping technicians often work outside, so they deal with different weather conditions and environments.
Average Salary: $51,940 (2024)
Number of Jobs: 57,400 (2023)
Future Job Outlook: 6 percent growth expected between 2023 and 2033

be divided into smaller sections. The small sections are for homes. Surveyors also identify areas that should not be developed. These include endangered habitats.

Most survey technicians work for companies that provide engineering and surveying services. Some work for governments. Others work for utility companies.

Survey techs work with a team in the field. This means they need good communication skills. They also need to pay attention to details. They must measure and record information accurately. Problem-solving skills are useful, too. Often, there is not a clear line of sight to take a measurement. The line might be blocked by trees. Or it could be over a hill. Techs have to decide how to solve these problems.

Being in good physical shape is also important. Survey techs sometimes walk over rough terrain. They also hike to sites while carrying equipment.

GETTING STARTED

Surveying and mapping techs need a high school diploma. Math skills are important. Survey tech Zachary Gooch says:

> *I wish I had paid more attention to math in school. We use a lot of trigonometry and geometry every day. . . . Back then, I never thought I would need to use that kind of math.*[5]

Learning about geographic information systems (GISs) is important. GIS software

Digital maps allow survey techs to quickly look up old data about an area.

is used to create maps. Techs use CAD systems as well. Many websites offer online courses to learn these programs.

Survey technicians do not need a certification. Most begin by working under a surveyor or **cartographer**. They start with simple tasks. These include placing wooden markers. They are used to mark points. Another task is entering data.

As survey techs gain experience, they are given more responsibilities. They might eventually help decide the best way to survey a site. Jack Edwards began as an assistant land surveyor. He says:

> *[The job] gives me a great opportunity to get to grips with the different types of equipment, learning the principles of how they work, their limitations, and what situations are most appropriate for what type of equipment.*[6]

Survey techs must know how to use different kinds of technology, such as computer programs, global positioning systems, and drones.

Certification is not necessary. But it can give people more opportunities. The National Society of Professional Surveyors offers a survey technician certificate. This shows employers that a tech knows the needed skills.

Surveying involves many specialties. Technicians may specialize in crime scene surveying. Another specialty is geodetic surveying. Geodetic surveys make adjustments based on Earth's curve. Techs use remote-sensing instruments to precisely measure large areas. Geodetic surveys create control survey points. These are marked with permanent metal markers. Control points are used as a starting point when measuring smaller areas of land.

ON THE JOB

A survey technician's day starts with a team briefing. The team then travels to a jobsite. Techs inspect the site for anything that could cause problems. This includes vegetation and obstructions. They search for old survey markers. Survey markers are objects that mark important survey points. Techs might look for control points, too. These are areas that are known for being accurately measured. They are the central points from which surrounding areas are measured.

Field work has challenges. Weather may affect equipment. Technicians often use all-terrain vehicles to reach sites.

Survey techs use a total station to take measurements. This is a precise instrument that measures distance. It also measures angles. Technicians use global positioning systems to find locations. Levels measure elevation. Laser scanners collect 3D information about structures or land features. Drones help collect information in hard-to-reach places.

After measurements are recorded, the team goes back to the office. Angles and elevations are calculated from survey data.

Mapping the Ocean

Danielle Power is a survey technician. She works for the National Oceanic and Atmospheric Administration. Power works aboard the ship *Pisces*. Her job is to collect information about the ocean. This includes temperature and depth. She then creates maps of the ocean. These maps display data about water chemistry and fish species at different depths.

Existing maps and surveys are updated with new data. Survey techs might work with a photogrammetrist. Photogrammetrists use satellite images to create maps. They also use images taken from the air. All this information is put into a GIS program. It updates maps. It also creates new ones.

Survey techs then plan for their next jobsite. They might be putting out markers for a new school. Or they might be laying out a park boundary. Gooch says, "That's what I like about the job. There is no normal day."[7]

FIND OUT MORE

Bureau of Land Management
www.blm.gov/programs
The Bureau of Land Management (BLM), which is part of the US Department of the Interior, provides a list of programs that people can get involved in to help protect the environment. These include land surveys.

National Society of Professional Surveyors
https://nsps.us.com
The National Society of Professional Surveyors (NSPS) site has information about becoming certified as a surveying technician. NSPS also offers educational resources for people interested in becoming a survey tech.

MACHINIST

Machinists use machine tools to make parts. These are made out of metal or other materials. Machinists use milling machines. Milling machines have rotating cutters. These remove metal from a part. Another common tool is a lathe. Lathes are used to shape cylinders. Machinists also work with drilling machines, lasers, and measuring tools.

Machinists can make anything from simple bolts or screws to more complicated parts that are used in cars and manufacturing machines.

Machinists make parts of all sizes. They make tiny screws that hold bones in place after surgery. Or they make parts for vehicle engines. They may make one item many times. Other times, they create a piece that is one of a kind. Orlando Morales is a machinist. He discovered the career when

Machinist

Minimum Education: High school diploma
Personal Qualities: Analytical and mechanical skills, good at working with hands, physically fit
Certification and Licensing: Not required, but certifications can help advancement
Working Conditions: Machinists work in machine shops and factories. They often handle dangerous tools such as milling machines, drilling machines, lathes, and lasers.
Average Salary: $57,700 (2024)
Number of Jobs: 357,000 (2023)
Future Job Outlook: Little to no change

When cutting parts, milling machines spray a liquid called coolant to prevent equipment from overheating or being damaged.

he toured a rice company. He explained, "I witnessed firsthand how the machinist cut up some metal and made a new part for the machine and got it back up and running in no time. And that was really amazing."[8]

A machinist creates parts designed by an engineer. These parts must be precise. Machinists must read blueprints. They need to understand how different machines work. And they must adapt to new technology.

Today, some machinists work with manual tools. Others work with computer numerical control (CNC) machines. CNC machines use software to automatically set the tasks of a machine. The software controls the speed and position of the

During apprenticeships, machinist apprentices learn how to operate CNC machines, read blueprints, use different tools, and more.

cutting tool. Working with CNC machines requires computer programming skills.

GETTING STARTED

Machinists need a high school diploma. Geometry and trigonometry classes are important to take. Some high schools offer drafting or metalworking classes. These skills are important for machinists.

Becoming a machinist requires on-the-job training. Some people work 12 to 18 months for a company to learn the job. Others may go to a technical school. These programs take about 2 years. Some organizations offer certifications for different machinist skills. One is the National Institute for Metalworking Skills.

Another path to a machinist career is becoming an apprentice. Apprentices learn a trade under an experienced professional. Apprenticeships usually last 4 years. Apprentices work 40 hours per week. They also take classes. These help them learn the skills needed to become machinists. Jaden Giles did an apprenticeship program. He says:

> *There's a lot of obstacles you run into, but that's the great thing about having a mentor that can . . . sit down and thoroughly help you, and you get . . . 100 percent hands-on attention.*[9]

Apprentices become journey-level machinists after their apprenticeship. This means they have the necessary skills to work on their own. They know how to work with different tools. Journey-level machinists can further their education by getting certifications.

ON THE JOB

Some machinists work in factories. Others work in smaller shops. Machinists can work at research labs or military bases as well. Some even work on ships. The workplace

3D Printing

Many companies use 3D printing to create parts. This is also known as additive manufacturing. Parts are built by adding thin layers of material instead of cutting it away. 3D printing takes time. It also costs more than old-style metalworking. But many companies use it to make prototypes and custom parts.

Machinists can get certified in many areas, including milling, grinding, and CNC programming.

is clean. It has good lighting. This helps the machinist meet certain standards.

Safety is important. Machinists wear earplugs when working with loud machines. They wear safety glasses, too. These protect their eyes.

Machinists must pay attention to avoid injury from the machines they use. They must also be aware of large machinery around them. These include cranes. Large machines often lift heavy items. These items can cause serious injury.

Machinists need to be careful when working with hazardous materials. They should wear protective suits and face shields. Gloves are important to wear when handling dangerous materials.

Many machinist jobs require workers to do Occupational Safety and Health Administration training to make sure they know proper safety techniques.

Machinists begin by carefully reading the instructions for making a part. They must decide exactly how to cut the material they are working with. They then set up the machine to make the part. While the machine is cutting, they monitor the process. Once the part is cut, machinists might have to grind it to smooth it out. Finally, they examine the products to make sure they are exactly right.

Machinists need to pay attention to details. This helps ensure parts are made correctly. They also need to have good problem-solving skills. If there is an issue with a part, machinists have to figure out how to fix it.

FIND OUT MORE

Fabricators and Manufacturers Association
www.fmamfg.org/training/certificates
The Fabricators and Manufacturers Association (FMA) offers certifications and technical training in metal manufacturing. This includes metal fabrication fundamentals training.

The Manufacturing Institute
www.themanufacturinginstitute.org/students
The Manufacturing Institute provides opportunities for young people to find out more about manufacturing careers.

OTHER JOBS IN THE ENGINEERING INDUSTRY

Civil Engineering Technician

Civil engineering technicians work with civil engineers to plan, design, and build large projects. These include hydropower plants. Civil engineering techs often begin their day with a site visit. They take measurements. They make sure sites and workers are following safety rules. And they ensure projects follow government regulations. Civil engineering techs also maintain the files and records to keep a project organized.

Electrical Engineering Technician

Electrical engineering technicians help engineers design and develop equipment powered by electricity. They assemble electrical systems. They build, test, and repair electrical instruments. Electrical techs find solutions to problems in a design, too. They test equipment and make adjustments if needed. Lathes and drills are just a few of the tools electrical techs use.

Chemical Engineering Technician

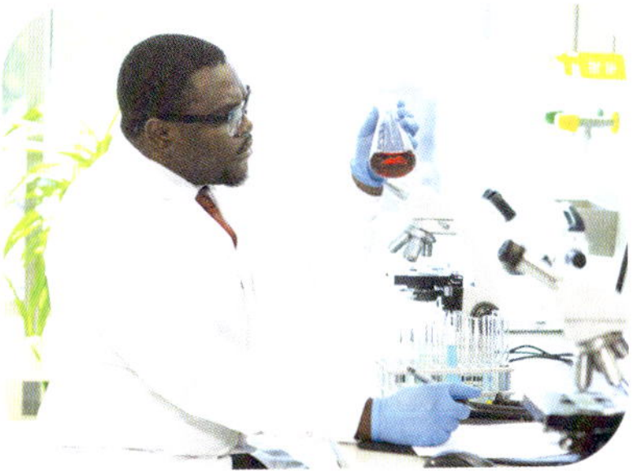

Chemical engineering technicians work in a laboratory. They help chemical engineers develop and test chemical products. They do tests to make sure chemical products meet government regulations. Chemical techs test products to determine what chemicals they contain. They also maintain and clean lab equipment. Most chemical techs do hands-on training.

Mechanical Engineering Technician

Mechanical engineering technicians help engineers design, test, and build machines. They might prepare designs using CAD software programs. They might also help make cost estimates based on designs. Their job is to troubleshoot new designs. Or they help improve old designs. Setting up and testing equipment is another part of a mechanical engineering technician's job.

GLOSSARY

cartographer

a professional mapmaker

contaminants

substances that poison or pollute something

elevation

the height of land above sea level

fruition

successfully completing something

hazardous

dangerous

prototype

the first model of a product or machine

specifications

detailed descriptions of designs and materials needed to make something

toxic

poisonous

zoning

dividing an area into sections for different uses

SOURCE NOTES

CHAPTER ONE: ENVIRONMENTAL ENGINEERING TECHNICIAN

1. Quoted in "Environmental Engineering Technician Reviews and Advice," *US News and World Report*, n.d. https://money.usnews.com.

2. Quoted in "Environmental Engineering Technician Reviews and Advice."

CHAPTER TWO: CAD TECHNICIAN

3. Quoted in "A Day in the Life of a Design Drafter," *McKee Foods Corporation*, March 22, 2023. https://jobs.mckeefoods.com.

4. Quoted in Elka Torpey, "You're a *What*? CAD Designer," *US Bureau of Labor Statistics*, May 2019. www.bls.gov.

CHAPTER THREE: SURVEYING AND MAPPING TECHNICIAN

5. Quoted in "Survey Technicians Aren't the Ones Who Ask Questions," *SkillPointe*, October 20, 2020. https://skillpointe.com.

6. Quoted in "A Day in the Life of an Assistant Land Surveyor," *RPS Group*, n.d. www.rpsgroup.com.

7. Quoted in "Survey Technicians Aren't the Ones Who Ask Questions."

CHAPTER FOUR: MACHINIST

8. Quoted in Don Sears, "They're Already Doing It: The Millennial Machinists Helping the Skills Shortage," *MSC*, May 22, 2018. www.mscdirect.com.

9. Quoted in "A Day in the Life of a Machinist Apprentice—JD Machine," *YouTube*, uploaded by Utah Department of Workforce Services, August 5, 2021. www.youtube.com.

INDEX

IMAGE CREDITS

Cover: © LightField Studios/Shutterstock Images
4: Red Line Editorial
5: Red Line Editorial
7: © pingpongcha_photo/Shutterstock Images
8: © Vastram/Shutterstock Images
11: © Kamonrat/Shutterstock Images
14: © Dejan Marjanovic/iStockphoto
17: © Chatchawal Phumkaew/Shutterstock Images
20: © SORN340 Studio Images/Shutterstock Images
23: © Chuck Rausin/Shutterstock Images
25: © AndreyPopov/iStockphoto
26: © SimoneN/Shutterstock Images
31: © Monkey Business Images/Shutterstock Images
35: © Juraj Kral/Shutterstock Images
39: © Kaspars Grinvalds/Shutterstock Images
40: © Westock Productions/Shutterstock Images
47: © tdub303/iStockphoto
49: © Pixel B/Shutterstock Images
50: © Monkey Business Images/Shutterstock Images
54: © Monkey Business Images/Shutterstock Images
55: © fotografixx/iStockphoto
58 (top): © Gorodenkoff/Shutterstock Images
58 (bottom): © Andrewshots/Shutterstock Images
59 (top): © Max Acronym/Shutterstock Images
59 (bottom): © Foto Eak/Shutterstock Images

ABOUT THE AUTHOR

Cynthia Kennedy Henzel has degrees in education and geography. She has written more than a hundred books for young people. These include fiction and nonfiction on subjects such as geography, science, culture, and history. Henzel enjoys discovering new things—especially discovering how something works or a new solution to a problem. But she isn't an engineer. She just likes writing about them.